Animal Habitats
The Best Home of All

by Nancy Pemberton
illustrated by Tom Dunnington

Created by

Distributed by CHILDRENS PRESS®
Chicago, Illinois

Grateful appreciation is expressed to Elizabeth Hammerman, Ed. D., Science Education Specialist, for her services as consultant.

Library of Congress Cataloging in Publication Data

Pemberton, Nancy.
    Animal habitats : the best home of all / by Nancy Pemberton ;
illustrated by Tom Dunnington ; created by Child's World.
     p. cm. — (Discovery world)
  Summary: A simple introduction to the different kinds of homes of
different kinds of animals.
  ISBN 0-89565-578-0
  1. Animals—Habitations—Juvenile literature. [1. Animals—
Habitations.   2. II fa04 02-20-90.]   I. Dunnington, Tom, ill.
II. Child's World (Firm)   III. Title.   IV. Series.
QL756.M364   1990
591.56'4—dc20   90-30633
                                        CIP
                                         AC

Animal Habitats
The Best Home of All

Where is Discovery World?
It can be anywhere—
anywhere at all.
It can be at a zoo,
a museum, or beside the sea.
It can be in a pond,
a library, or up in a tree.
All you need is curiosity.
It can be right at
your own back door.
All you need to do is EXPLORE!

So come along and find out more about . . .

ANIMAL HABITATS!

There once was a little polar bear who lived near the top of the world, in the icy Arctic.

One day she said to a snowy owl, "I am
tired of living in a white world of ice and
snow. I am going to travel the world to
find a better home." And off she went.

But when she came to the edge of the
ice, the sea looked so very big. "Oh my,"
she said, "I have a long way to go. Maybe
I should have dinner first."

So she ate and ate until her belly was
full. "I'm too full to go now," she said.
"Maybe I need a little nap before my trip."
In a moment she was fast asleep.

Suddenly she heard noises she had
never heard before. When she looked up,
she saw a monkey, hanging from a tree.
"Where am I?" she asked.

"You are in my home," said the
monkey. "This is a tropical rain forest. I
live here with colorful birds and giant
spiders and snakes.

"There are lots of vines to swing from
and fruits to eat. This is the best home of
all."

Just then it began to rain. Down it
poured. "This home may be best for you,"
said the bear, "but not for me. It is too
dark and rainy." And off she went.

Soon she found that she was surround-
ed by hot, dry sand and prickly cactus
plants. "Where am I?" asked the little bear.
"You are in my home," said a lizard.

"I live in the desert. Scorpions and
snakes are my neighbors. In the daytime
the sun is very hot.

"Some of us hide in the shade or underground to beat the heat. When night-time comes, the desert cools off. I think the desert is the best home of all."

"This home may be best for you," said
the little bear, "but it is much too hot for
me. Besides, I'm thirsty!" And off she
went.

Suddenly she found herself at the edge of a great grassy plain. "Where am I?" asked the little bear.

"You are in my home," said an elephant.

"These are the grasslands. I live here
with giraffes and zebras and the fast-
running ostrich.

"There is lots of room to run and grass to eat. This is the best home of all."

Just then, the little bear heard a terrible ROAR! "Wh-what was that?" she asked.

"That was just my neighbor, the lion,"
said the elephant.

"This home may be best for you," said
the bear, "but not for me. I wouldn't want
a lion for a neighbor. Besides, I don't like
to eat grass." And off she went.

In no time, the little bear found that she
was high on a mountaintop. "Where am
I?" she asked.

"You are in my home," said a mountain
goat.

"I live in the mountains. Bighorn sheep
and cougars live here too. We are all very
good climbers.

"It is so cold up here that trees cannot grow. But the view is great! This is the best home of all."

"It may be the best home for you," said the little bear, "but not for me. I am afraid of heights!

"Oh, how I wish I were back in the white world of ice and snow. It's not too wet or too dry or too hot or too high. And it's full of the food I love to eat!"

Suddenly, she felt something cold and
wet on her nose. She opened her eyes.
Snow and ice were everywhere!

"I must have been dreaming!" cried the little bear. "Hooray! This is just where I want to be. The Arctic is the best home of all—for a polar bear like me!"

MORE TO EXPLORE

Professor Facto says:

The best home of all for an animal is a place that has what the animal needs to live, such as the right food. Every animal has his own special needs.

Beavers need to live near water and trees. They eat tree bark, and they use trees to make their lodges. Beavers are good swimmers. They build their lodges in water. The best home of all for a beaver is a lake or stream with trees around it.

Giant pandas eat mostly bamboo plants. The best home of all for a giant panda is in a bamboo forest in China.

Koala (koh-AH-luh) bears eat mostly the leaves of eucalyptus (yoo-kuh-LIP-tus) trees. They spend most of their time up in the trees. They come down only to move from one tree to another. The best home of all for a koala bear is in Australia, where lots of eucalyptus trees grow.

Polar bears hunt seals for food. The best home of all for polar bears is the Arctic, where there are plenty of seals.

Another word for an animal's home is **habitat.** An animal's habitat is the area where he lives. What is your habitat?

Be an Animal Detective!

Explore your backyard or a nearby park. Look for animals. Find a good hiding place, and stay very quiet. Watch the animals in their habitat. What do they eat? Do the animals have fur, feathers, or scales? Do they make any sounds? How do they move?

Just Imagine

If you could be an animal—other than a human—what kind would you want to be? Why? Where would you live? Draw a picture of that animal in his habitat.

Make a Habitat

You can make a home for worms. Here's how.

You will need:

—a large glass jar (1-2 gallons or 3-6 liters) with a large mouth
—good, rich soil
—gravel
—food for worms (small pieces of lettuce, cornmeal)
—earthworms

1. Fill the bottom of the jar with a layer of gravel mixed with soil. Add loose soil to fill the jar ⅔ full. Keep the soil moist.
2. Add pieces of lettuce and a handful of cornmeal on top of the soil.
3. Dig up several earthworms and move them to their new home.
4. Tape black paper to the sides of the jar for one week. That makes it more likely that the worms will tunnel near the glass.

What do you notice when you take off the black paper? Watch how the worms move and eat.

When you are through observing the worms, return them to their habitat outdoors.

INDEX